Everyday
Mathematics®

The University of Chicago School Mathematics Project

Mathematics at Home
Book 2

 **Education**

everyday**math**.com

 **Education**

Printed in Mexico

Send all inquiries to:
McGraw-Hill Education
P.O. Box 812960
Chicago, IL 60681

ISBN   978-0-07-657578-7
MHID 0-07-657578-0

2 3 4 5 6 7 8 9 DRN 16 15 14 13 12 11

Authors
Ann E. Audrain
Jean Bell
Jeanine O'Nan Brownell
Patrick Carroll
Deborah Arron Leslie

Consultant
Max Bell

Third Edition Early Childhood Team Leaders
Deborah Arron Leslie, David W. Beer

Technical Art
Diana Barrie

Contributors
Dorothy Freedman, Nancy Hanvey, Sue Lindsley, Ellen Ryan

Photo Credits
Cover Sharon Hoogstraten/Courtesy of Dave Wyman; **Back Cover** Dorling Kindersley/Getty Images.

The **McGraw·Hill** Companies

Contents

Introduction

Learning Mathematics

This is the second of three *Mathematics at Home* books for *Pre-Kindergarten Everyday Mathematics*. Enjoy these new activities with your child, but continue to use favorite activities from Book 1.

As with Book 1, you are free to vary the numbers or themes in any of the activities. Always remember to let mutual enjoyment set the tone and pace. The idea is to explore mathematics with your child in natural and playful ways.

Browse through the list of related children's books provided in the back. Perhaps you will find some of them on your next trip to the library. These books, and others like them, can spark mathematical thinking and discussions with your child.

Grocery Store Mathematics

Shopping List

Plan a trip to the grocery store together. To start, make a list of things you need, such as 2 cans of soup, 1 loaf of bread, 4 lemons, and so on. As you shop, discuss where you will find the items on your list.

Numbers All Around

Look at the aisle numbers as you shop. What do they sell in Aisle 3? Where do you think Aisle 4 is? Also look for other numbers in the store, such as prices, sizes, and weights. Talk about what information the different numbers provide.

Counting Fun

Count how many different types of pasta you can find. How many eggs are in a carton? How many juice boxes are in a package? Look around for other things to count.

Which Is Biggest?

Look for items in the store that come in different sizes. Which is the biggest can of peaches? Which is the smallest box of cereal? How many different sizes of milk containers do you see?

Shape Search

Look for different shapes throughout the store. Can you find items that remind you of circles, ovals, squares, rectangles, and triangles? They might not be the exact geometric shapes, but that's okay.

Using the Scale

Look for a scale in the produce department. Weigh a bag of apples and a bag of oranges. Which one weighs more? What else can you weigh on the scale?

Checkout Sort

Sort the items in the cart. You might group together the things that are cold or frozen and things that need to be packed carefully, such as eggs or fruit. Or, sort using your own categories.

Bagging and Unpacking

Try counting how many items go into one of the bags at the checkout counter. When you get home, unpack that bag, counting the items as you go. Was it the same number?

Sorting

Recycling Sort

If your community has a recycling program, you might sort newspaper, cans, and plastic bottles into separate bins. Be careful of any sharp edges!

Sorting in the Kitchen

Together, sort the items in the kitchen cabinet or pantry. You can sort by categories, such as canned food, boxed food, and bagged food. Or, use your own categories for sorting. Fresh fruit can also be sorted and grouped by type.

After washing dishes, you can help organize the silverware drawer by putting all the forks in one section and all the spoons in another.

Laundry Time

Help sort the laundry by color or other categories: for example, school and play clothes, or child and adult clothes. When the wash is done, make piles of folded clothes for different family members. Can you match socks into pairs?

Mail Delivery

Make pretend mail by writing family members' names on several envelopes, or use real mail. Sort the mail by the names on the envelopes and then deliver it.

Clean-up Categories

Help to clean a room by sorting toys, clothes, books, and other items into categories. You might use baskets or shoe boxes for different types of toys, such as toy cars, doll clothes, or stuffed animals.

Patterns

Musical Patterns

Listen to a sound pattern and then repeat it.

    Clap! Clap! Tap!

    Clap! Clap! Tap!

Take turns making sound patterns using your feet and hands.

Play some music and listen to the beat. Now clap to the beat. Do you think that's a pattern? Try copying the pattern another way, such as with a snap or a stomp.

Movement Patterns

Create a pattern as you walk together: step, step, hop; step, step, hop; step, step, hop… .

Patterns around the House

Try arranging red and black checkers into a color pattern. You can also make patterns with different-shape crackers or pasta, colored cereal, or two different sizes of cans. If the cereal or pasta has holes, use yarn to string them in a pattern for a necklace or bracelet.

Arrange your toy cars, stuffed animals, or other toys into a pattern. Ask a family member to guess what comes next.

Patterns in Books

Read a book with patterned language—words or phrases that repeat or rhyme. See page 14 for some suggestions. As you read together, pause and see whether you can guess what comes next. Also look for books with patterned borders, such as shapes or objects that repeat around the edge of a page.

Position

Over and Under

Make a tower of different-color blocks or of different types of canned food items. Which one is on the *top?* Describe the one on the *bottom.* Is the green block *over* or *under* the red block?

Traveling Directions

When walking or traveling in a car, bus, or train, use position words to describe the things you see. For example, "We are *next to* the river now." "There's a huge tree *above* us!" or "Did we just go *under* a bridge?"

Look for familiar sights on the way to places you go frequently. Point these out as you pass them. For example, "We're going *over* the big hill. That means we're almost at Grandpa's house!"

Toy Hunt

Hide a toy while a family member closes his or her eyes. Give clues using position words, such as, "It is *on* the bookshelf," or "It is *next to* the chair."

You can also play I Spy using position words. Give clues such as, "I spy, with my little eye, something that is *behind* the lamp."

It's next to the chair...

Mathematics Note

Initially, children will find it easier to respond to clues rather than provide them. So you should be the one to begin giving the clues. After listening to your clues for some time, your child will be ready to give clues, too.

Position Books

Read a book together that features position and spatial relationship language. Discuss what those words mean in the story. See page 14 for suggestions.

More Counting

Step Counts

Count how many steps it takes to walk from the front door to the kitchen. Now try from the front door to your bedroom. Which took more steps? Which is farther? Does the number change if you use giant steps? What about tiny baby steps?

Counting around Town

Together, count the number of floors in a building, cars in a passing train, cars lined up at a traffic light, ball bounces, jump rope turns, and so on.

Countdowns and "Count Ups"

Count during "waiting" times, such as waiting for a bathtub to fill or waiting for a turn with a toy. You might count up to see how quickly you can clean up your toys. Or, you might have someone do a countdown *(10, 9, 8, 7, 6, 5, 4, 3, 2, 1, Blast off!)* to see whether you can get your jacket or shoes on before "Blast off!"

Look and Count

How many windows are there in your home, or in one or two rooms of your home? How many doors are there? Are there more windows or doors? You can also count other things, such as faucets, lamps, or mirrors. It might be fun to guess how many, then count.

Some Books for Children

Patterns and Sorting

Giganti, Jr., Paul, *How Many Snails? A Counting Book* (Greenwillow, 1988)

Harris, Trudy, *Pattern Bugs* and *Pattern Fish* (Lerner Publishing Group, 2001)

Swinburne, Stephen, R., *Lots and Lots of Zebra Stripes: Patterns in Nature* (Boyds Mills Press, 1998)

Reid, Margarette, S., *The Button Box* (Dutton Juvenile, 1990)

Patterned Language

Martin, Bill, Jr., *Brown Bear, Brown Bear, What Do You See?* (Henry Holt and Co., 1992)

Taback, Simms, *There Was an Old Lady Who Swallowed a Fly* (Viking Juvenile, 1997)

Williams, Sue, *I Went Walking* (Gulliver Books, 1990)

Position and Spatial Relationships

Carle, Eric, *The Secret Birthday Message* (HarperCollins, 1972)

Fanelli, Sara, *My Map Book* (HarperCollins, 1995)

Hoban, Tana, *Over, Under and Through* (Simon & Schuster Children's Publishing, 1973)

Hutchins, Pat, *Rosie's Walk* (Little Simon, 1998)

Rosen, Michael, *We're Going on a Bear Hunt* (Margaret K. McElderry, 1989)

Counting

Bajaj, Varsha, *How Many Kisses Do You Want Tonight?* (Little, Brown, 2004)

Falwell, Cathryn, *Feast For 10* (Clarion Books, 1993)

Walsh, Ellen Stoll, *Mouse Count* (Harcourt Children's Books, 1991)